YOUR KNOWLEDGE HAS VALUE

- We will publish your bachelor's and master's thesis, essays and papers

- Your own eBook and book -
  sold worldwide in all relevant shops

- Earn money with each sale

Upload your text at www.GRIN.com
and publish for free

Bibliographic information published by the German National Library:

The German National Library lists this publication in the National Bibliography; detailed bibliographic data are available on the Internet at http://dnb.dnb.de .

Imprint:

Copyright © 2017 GRIN Verlag, Open Publishing GmbH
Print and binding: Books on Demand GmbH, Norderstedt Germany
ISBN: 9783668538153

This book at GRIN:

http://www.grin.com/en/e-book/376165/an-overview-of-phase-change-isotherm-and-isotherm-migration-method

Khairuzzaman Mamun

An Overview of Phase Change, Isotherm, and Isotherm Migration Method

GRIN Publishing

Iwate University

<u>Prepared by:</u>
Khairuzzaman Mamun

Submission Date: September 01, 2017

An Overview of Phase Change, Isotherm, and Isotherm Migration Method

Contents

Introduction:

This article is aimed at presenting the definition and mathematical discussion of some important thermodynamic terms such as phase change, isotherm and isotherm migration method etc. It overviews some theoretical literature on these topics and presents their interpretation. Phase-change, Isotherm, Stefan Problems, and Isotherm Migration method are described with graphical or mathematical representation to clarify the concepts so that one can get brief idea about them.

Phase-change:

The term phase transition or phase change is most commonly used to describe transitions between solid, liquid and gaseous states of matter, and, in rare cases, plasma. A phase of a thermodynamic system and the states of matter has uniform physical properties. During a phase transition of a given medium certain properties of the medium change, often discontinuously, as a result of the change of some external condition, such as temperature, pressure, or others. For example, a liquid may become gas upon heating to the boiling point, resulting in an abrupt change in volume. The measurement of the external conditions at which the transformation occurs is termed the phase transition. Phase transitions are common in nature and used today in many technologies.

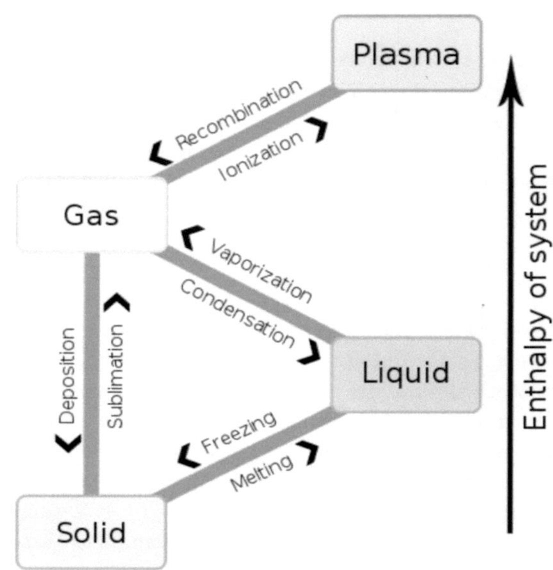

https://en.wikipedia.org/wiki/Phase_transition#/media/File:Phase_change_-_en.svg

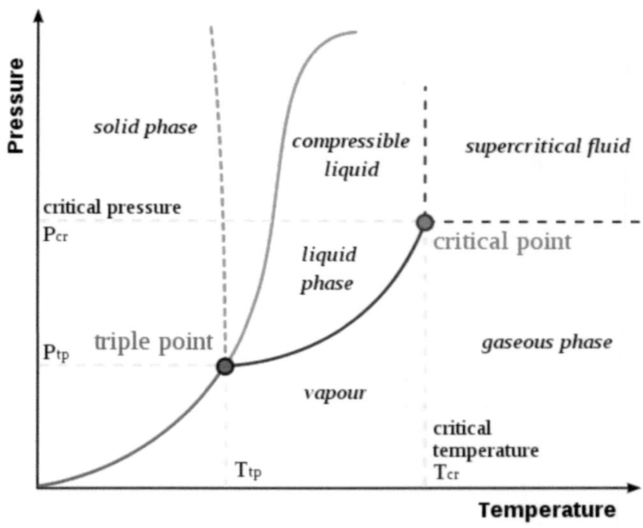

https://upload.wikimedia.org/wikipedia/commons/3/34/Phase-diag2.svg

During a change in state the heat energy is used to change the bonding between the molecules. In the case of melting, added energy is used to break the bonds between the molecules. In the case of freezing, energy is subtracted as the molecules bond to one another. These energy exchanges are not changes in kinetic energy. They are changes in bonding energy between the molecules.

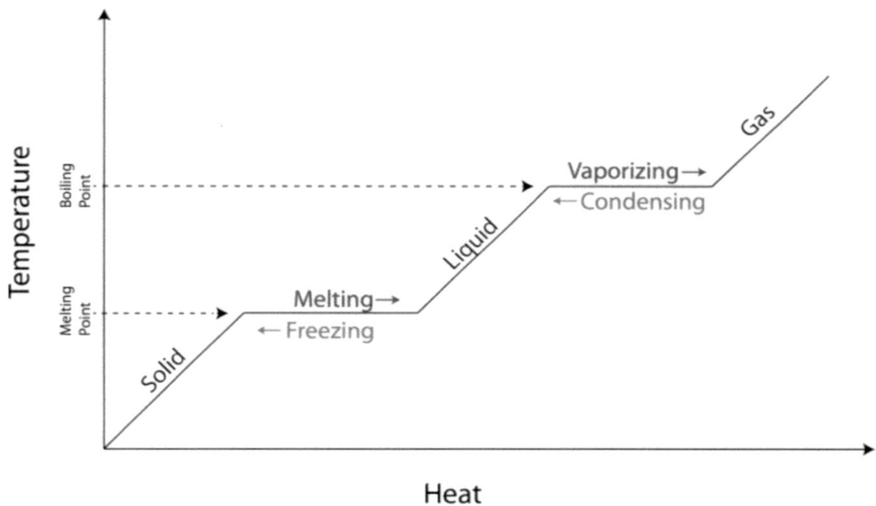

http://www.aplusphysics.com/courses/honors/thermo/phase_changes.html

If heat is coming into a substance during a phase change, then this energy is used to break the bonds between the molecules of the substance. The example we will use here is ice melting into water. Immediately after the molecular bonds in the ice are broken the molecules are moving (vibrating) at the same average speed as before, so their average kinetic energy remains the same, and, thus, their Kelvin temperature remains the same.

Isotherm:

The term Isotherm is derived from the Greek words, isos and thermē, the former meaning the equal and the latter heat. Isotherm thus means keeping the same temperature at a given time or on average over a given period.

Stefan Problems:

In mathematics and its applications, particularly to phase transitions in matter, a Stefan problem is a particular kind of boundary value problem for a partial differential equation (PDE), adapted to the case in which a phase boundary can move with time. The classical Stefan problem aims to describe the temperature distribution in a homogeneous medium undergoing a phase change, for example ice passing to water: this is accomplished by solving the heat equation imposing the initial temperature distribution on the whole medium, and a particular boundary condition, the Stefan condition, on the evolving boundary between its two phases. Note that this evolving boundary is an unknown surface: hence, Stefan problems are examples of free boundary problems.

A common-place physical phenomenon is the melting of a block of ice by raising its surface to a temperature above 0 degree C. The two phases, water and ice, are separated by a boundary on which melting occurs at 0 degree C and which moves further into the ice as time progresses. In the mathematical treatment, the motion of the boundary has to be determined and the usual equations of heat flow solved in the water and the ice. The solutions and the boundary movement are dependent on each other. The melting of ice is just one example of a whole class of problems commonly referred to as Stefan Problems. They include the propagation of phase changes in metals diffusion with absorption and processes controlled by discontinuous diffusion coefficients.

Methods of Solution:

In some cases analytical solutions can be obtained, according to Carslaw and Jaeger [1]. Several numerical methods, all based on finite-difference replacements of the original partial differential equation, differ in the way they cope with the movement of the boundary. As usual, in a one dimensional problem the region is covered by a grid of equally spaced lines. The various numerical methods have really explored all possible ways of using the grid. Special finite-difference formulas based on Lagrangian interpolation formula for unequal intervals have been used in the neighborhood of the boundary when it falls between two grid lines by Crank [2]. Unequal time intervals have been used, calculated so that the boundary moves always from one grid time to the next in one time step by [3]. The grid has been deformed so that the number of space intervals between the outer surface and the moving boundary remains constant, with suitable transformation of the basis equation by Murray and Landis [4].

Another method employs an apparent specific heat modified to include the latent heat in the appropriate region, according to Albasiny [5]. Finally, the whole grid has been moved with the velocity of the moving boundary in a method incorporating interpolating splines by Crank and Gupta [6]. Recently a novel way of handling heat flow problems has been proposed which is especially useful for tracking a moving boundary that occurs at a fixed temperature. In the usual heat flow equation in one dimension the temperature is expressed as a function of the independent space variable x, and time t i.e. u =u(x, t).

An alternative, however, is to seek a solution in which x is expressed as a function of u and t i.e. x =x (u, t) so that x becomes the dependent variable. We calculate the positions of a given temperature, which is of an isotherm at known times. Hence, the method is known as the Isotherm Migration Method (IMM) [7]. Philip [8] dealt with a problem in concentration dependent diffusion by making concentration an independent variable but he did not transform the diffusion equation in the same way as Dix and Cizek. Rose [9] derived a related transformed equation but did not develop a numerical method. The idea of tracking the moving isotherms in media with phase changes was published by Chernoua'ko [10] in 1969, though the English version appeared only in 1970 [11].

Melting Ice:

Suppose a plane sheet of ice initially occupies the region $0 \leq x \leq a$ and is being melted by the application of a constant temperature, u_0, on the surface $x = 0$. At any time, t, let the moving boundary separating the water from the ice be at $u_0(t)$. The region then consists of water with specific heat, $0 \leq x \leq x_0(t)$ density and thermal conductivity denoted by ρ, C and K respectively. The temperature, u, of the water satisfies the heat flow equation

$$\frac{\partial u}{\partial t} = k \frac{\partial^2 u}{\partial x^2} \tag{1}$$

Where $k = {}^{K}/_{C\rho}$, the heat diffusivity.

We take the ice to be initially at 0°C throughout. Otherwise we should have an equation similar to (1) for the temperature in the ice phase and containing appropriate heat parameters.

At the melting boundary, $x_0(t)$, the heat flowing per unit area from the water into the ice in a short time, δt, is $-(\frac{K\partial u}{\partial x})\delta t$. If the boundary moves a distance δx_0 in time δt, the heat required to melt the mass $\rho\delta x_0$ of ice per unit area is $L\rho\delta x_0$ where L is the latent heat of fusion for ice.

Equating these two amounts of heat and proceeding to the limit $\delta t = 0$, we see that the first condition to be satisfied on the boundary is

$$L\rho \frac{dx_0}{dt} = -K \frac{\partial u}{\partial x} \tag{2}$$

A second condition, since ice melts at 0°C is

$$u = 0, \ x = x_0, \ t \geq 0 \tag{3}$$

Commonly used variables

$$X = \frac{x}{a}, \ T = \frac{kt}{a^2}, \ X_0 = \frac{x_0}{a}, \ S = \frac{L}{o} \tag{4}$$

Lead to the following system of equations

$$\frac{\partial u}{\partial T} = \frac{\partial^2 u}{\partial X^2}, \quad 0 < x < x_0, \quad T \geq 0 \tag{5}$$

$$S\frac{dx_0}{dT} = -\frac{\partial u}{\partial X}, \quad X = X_0, \quad T \geq 0 \tag{6}$$

$$u = 0, \ X = X_0, \ T \geq 0 \tag{7}$$

$$u = u_0, \ X = 0, \ T \geq 0 \tag{8}$$

$$u = 0, \ \ 0 < X < 1, \quad T = 0 \tag{9}$$

Isotherm Migration Equations:

We wish to re-write the heat flow equation (5) so that X is expressed as a function of u and T. Since the temperature u is constant along an isotherm, we have-

$$du = \left(\frac{\partial u}{\partial X}\right) dX + \left(\frac{\partial u}{\partial T}\right) dT = 0$$

$$\Rightarrow \frac{\partial X}{\partial T} = -\frac{\left(\frac{\partial u}{\partial T}\right)}{\left(\frac{\partial u}{\partial X}\right)} = -\left(\frac{\partial u}{\partial T}\right)\left(\frac{\partial X}{\partial u}\right) \tag{10}$$

Substituting (5) in (10) and dropping suffices we obtain

$$\frac{\partial X}{\partial T} = -\left(\frac{\partial^2 u}{\partial X^2}\right)\left(\frac{\partial X}{\partial u}\right) \tag{11}$$

Now

$$\frac{\partial^2 u}{\partial X^2} = \frac{\partial}{\partial X}\left(\frac{\partial X}{\partial u}\right)^{-1} = -\frac{\partial^2 X}{\partial u^2}\left(\frac{\partial X}{\partial u}\right)^{-3} \tag{12}$$

Therefore,

$$\frac{\partial X}{\partial T} = \left(\frac{\partial X}{\partial u}\right)^{-2}\frac{\partial^2 X}{\partial u^2} \tag{13}$$

This equation represents X as a function of u and T.

The other equations (6) - (8) become

$$s\frac{dx_0}{dT} = -\left(\frac{\partial x}{\partial u}\right)^{-1}, \quad u = 0,\ T \geq 0 \tag{14}$$

$$u = 0,\ X = X_0,\ T \geq 0 \tag{15}$$

$$U = U_0,\ X = 0,\ T \geq 0 \tag{16}$$

We can approximate the derivatives in (13) by finite difference, in; the usual way and obtain an explicit expression for X_i^{n+1} , the value of X at $u = i\delta u$, $T = (n+1)\delta T$ in terms of values already available at $(i\delta u, n\delta T)$.

We find

$$X_i^{n+1} = X_i^n + 4\delta t\left\{\frac{X_{i+1}^n - 2X_i^n + X_{i-1}^n}{\left(X_{i-1}^n - X_{i+1}^n\right)^2}\right\} \tag{17}$$

The corresponding finite-difference replacement of (14) is

$$(X_0)^{n+1} = (X_0)^n - \frac{\delta T \delta u}{s(X_0^n - X_1^n)} \tag{18}$$

A rigorous analysis of the stability of the non-linear finite difference scheme has not been attempted. Dix and Cizek [7] consider instead the coefficient of X_i^n in (14). In order that the isotherms should move with time in the manner expected X_i^{n+1} should increase with X_i^n i.e. the coefficient of X_i^n must be positive. This leads to the criterion

$$\partial T < 1/8(X_{i-1} - X_{i+1})^2 \tag{19}$$

Which allows ∂t to change as the solution proceeds. The truncation error [7] of the IMM is proportional to Δt and $(Au)^2$.

Conclusion:

Phase change, isotherm and isotherm migration method are very important terms to learn in thermodynamics. Therefore, I have tried to give a short overview on the Phase change, isotherm and isotherm migration method. From this article one can get brief idea about the phase change and isotherm migration method.

Acknowledgement:

From the core of my heart I would like to convey my earnest admiration, loyalty and reverence to the almighty Allah, the most merciful, for keeping everything in order and enabling me to complete this article successfully. Then let me express my profound gratitude to my tutor for his inspiration. It would be impossible for me to finish this article without his enthusiasm and motivation. It has also been a pleasure to have worked with him.

References:

[1]   Carslaw, H.S., and Jaeger, J.C. Conduction of Heat in Solids, O.U.P., 1959, Chap.XI .

[2]   Crank J., Quart J., Mech App.Math., 10 (1957) 220.

[3]   Douglas J., and Gallie T.M., Duke Math. J., 22 (1955) 557.

[4]   Murray W.D., and Landis F. Trana. ASME 81 (1959) 106.

[5]   Albasiny, E.L.: Proc.I.E.E. 103 (1956) Series B, Parts 1-3, 158.

[6]   Crank,J., and Gupta,R.S., JIMA (in the press)

[7]   Bix, R.C., and Cizek, J. Heat Transfer 1970, Vol.1, 4th International Heat Transfer Conference, Paris Versailles, Elsevier, 1970.

[8]   Philip, J.R. Trans.Faraday Soo 51 (1955) 885.

[9]   Rose, M.E. SIAM J.of Applied Maths.15 (1967) 495.

[10]   Chernous'ko, F.I, Zh.Prikl.Mekh.i,Tekhn. Fis.No.2 (1969) 6.

[11]   Chernoua'ko, F.L. International ChenuEng.10, No.1, (l970) 42.

[12]   Goodman, T.R Adavances-5 in Heat Transfer, Vol.1 Academic Press, New York, 1964.

[13]   Coldrey, T.J. M.Tech Dissertation, Brunel University, 1966.

YOUR KNOWLEDGE HAS VALUE

- We will publish your bachelor's and master's thesis, essays and papers

- Your own eBook and book - sold worldwide in all relevant shops

- Earn money with each sale

Upload your text at www.GRIN.com and publish for free